TINY

The Invisible World of Microbes

ILLUSTRATED BY EMILY SUTTON

WALKER BOOKS
AND SUBSIDIARIES

LONDON · BOSTON · SYDNEY · AUCKLAND

You know about big animals

and you know about small animals ...

but do you know that there are creatures so tiny
that millions could fit on this ant's antenna?

So tiny that we'd have to make the ant's antenna
as big as a whale to show them to you?

ANTENNA

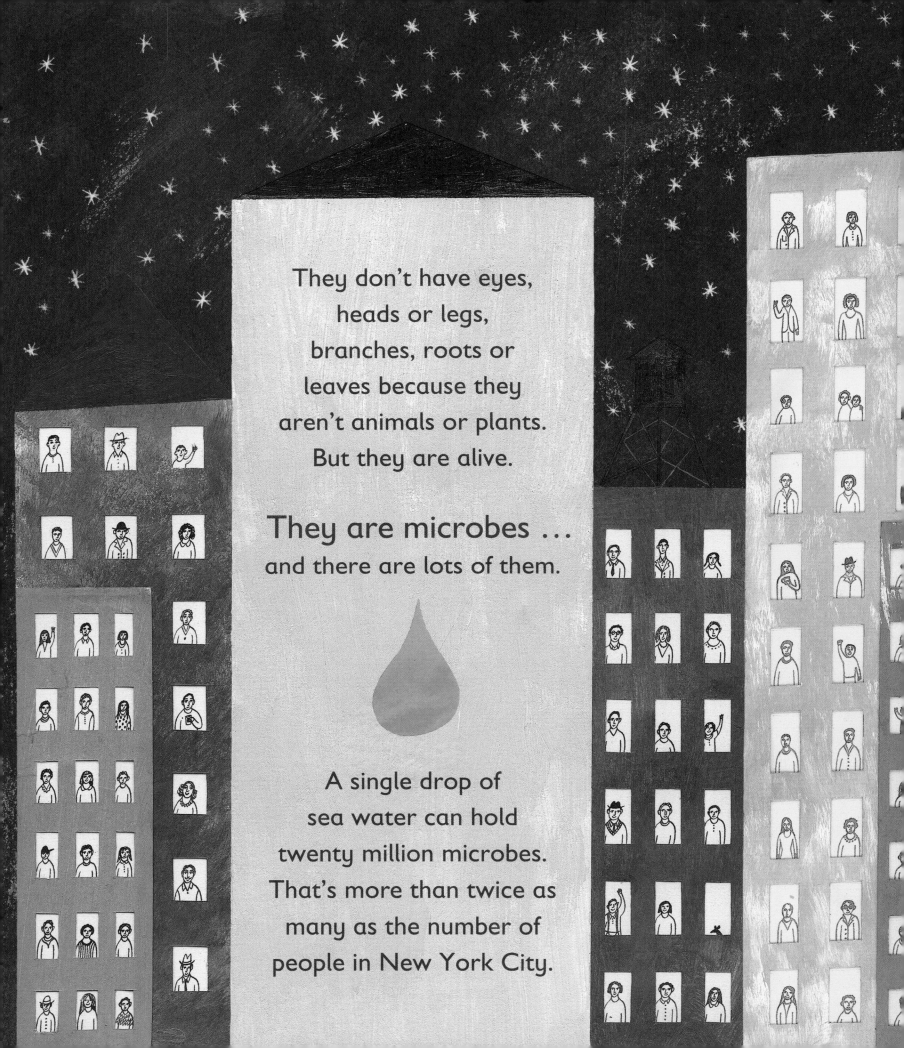

They don't have eyes,
heads or legs,
branches, roots or
leaves because they
aren't animals or plants.
But they are alive.

They are microbes …
and there are lots of them.

A single drop of
sea water can hold
twenty million microbes.
That's more than twice as
many as the number of
people in New York City.

And a teaspoon of soil can have as many as **a billion** microbes. That's about the same as the number of people in the whole of India.

LIKE VOLCANOES

OR INSIDE ROCKS

OR AT THE BACK OF YOUR FRIDGE

Microbes live everywhere — in the sea, on land, in the soil and the air. They live in places where nothing else does, like volcanoes, or inside rocks, or at the back of your fridge,

and also on the outside and inside of plants and animals.

Right now there are more microbes living on your skin than there are people on Earth, and there are ten or even a hundred times as many as that in your tummy.

(Don't worry, although microbes sometimes make you sick, the ones that live in you and on you all the time help to keep you well.)

But even though they are so small, they are not all the same.
Some are tinier than others, as different in size as ants and whales.
There are lots of different kinds, more than there are
different kinds of animals and plants.

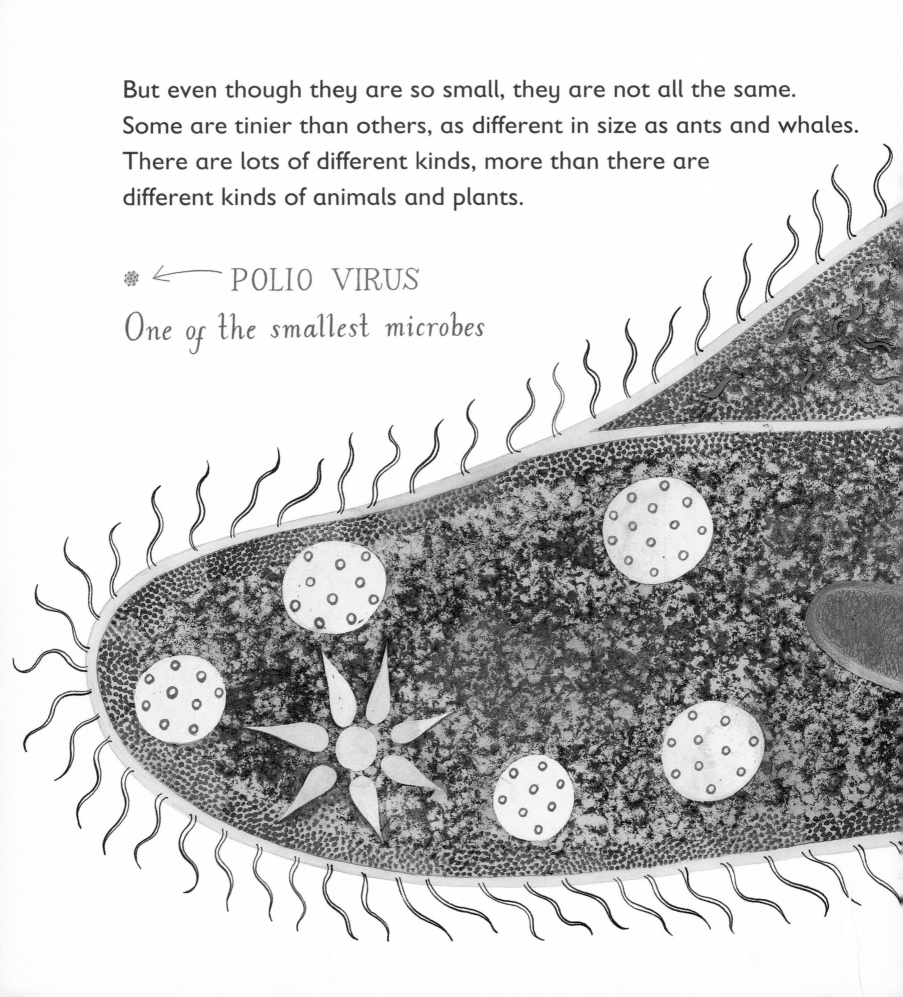

← POLIO VIRUS
One of the smallest microbes

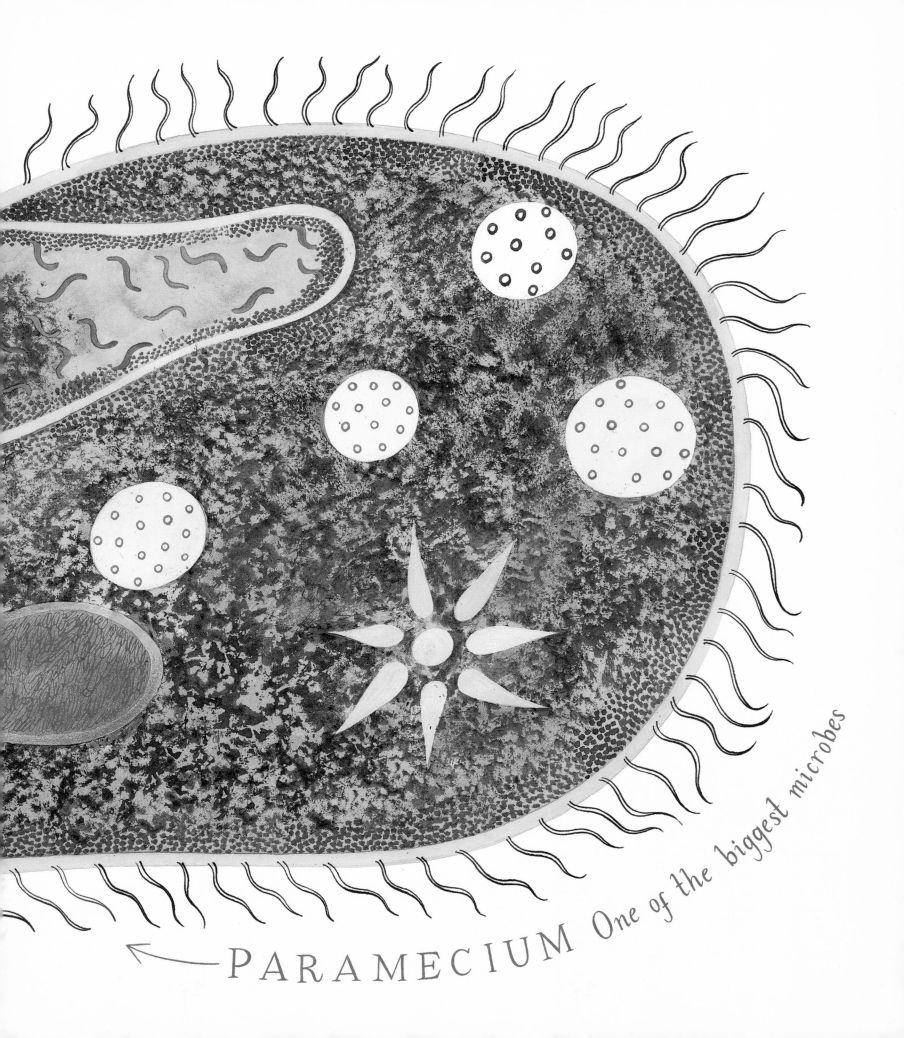

PARAMECIUM One of the biggest microbes

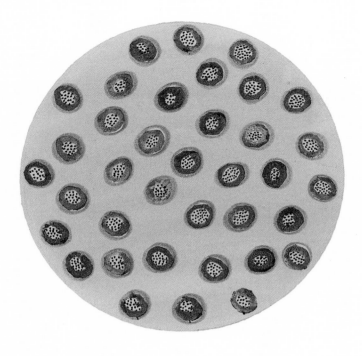

Some microbes are round

Some are skinny

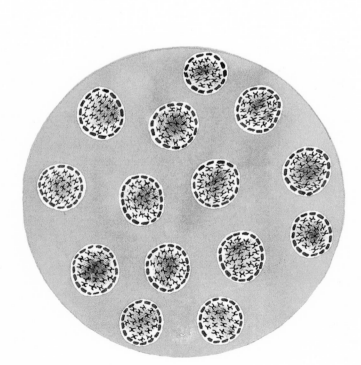

Some look like shells

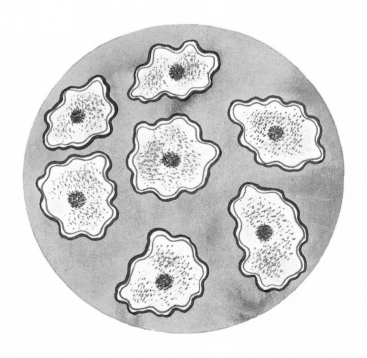

Some are squishy

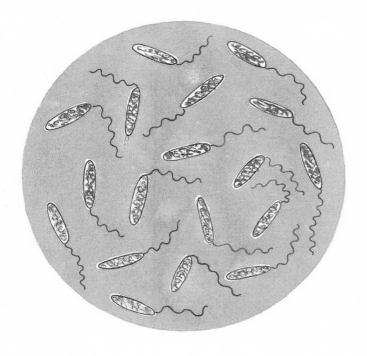

Some have wiggling tails

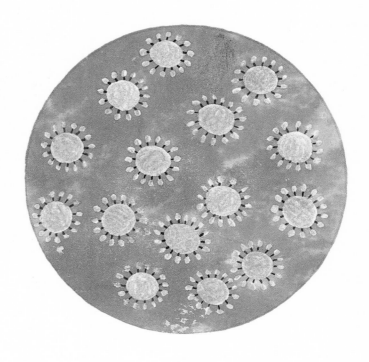

Some look like daisies

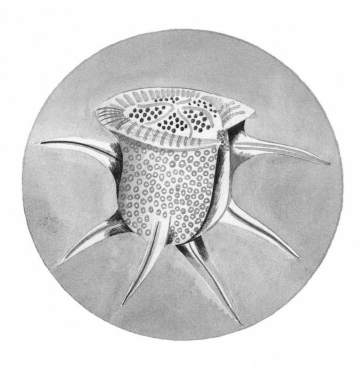

Some look like spaceships

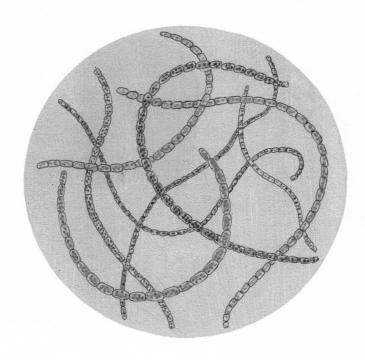

Some look like necklaces

They can eat anything: plants, animals (alive or dead), even oil and rocks. They're too small to have mouths so they just soak up what they need through their skin.

MENU

PLANTS

ANIMALS

ROCKS

OIL

That's why the things microbes eat
don't disappear in bites. They change,
slowly, into something else...

FOOD *INTO* COMPOST

MILK *INTO* YOGHURT

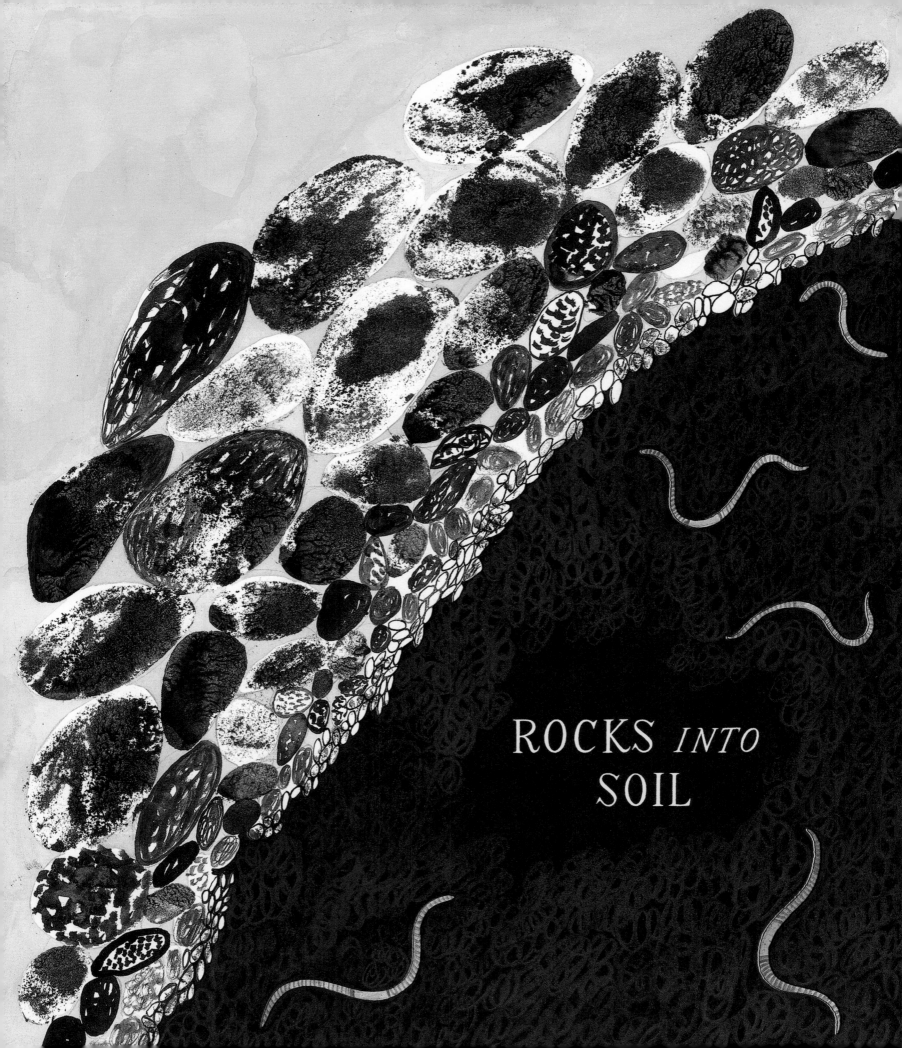

ROCKS *INTO*
SOIL

And when microbes are well fed they are really,
really good at making more microbes.
They simply split, so that where there was one,

twenty minutes later there are two …

and then four,

and then eight,

and then sixteen.

Starting with just one microbe,
which of course would be far too small for you
to see (one called E. coli for example), it would
take eleven and a half hours for there to be enough

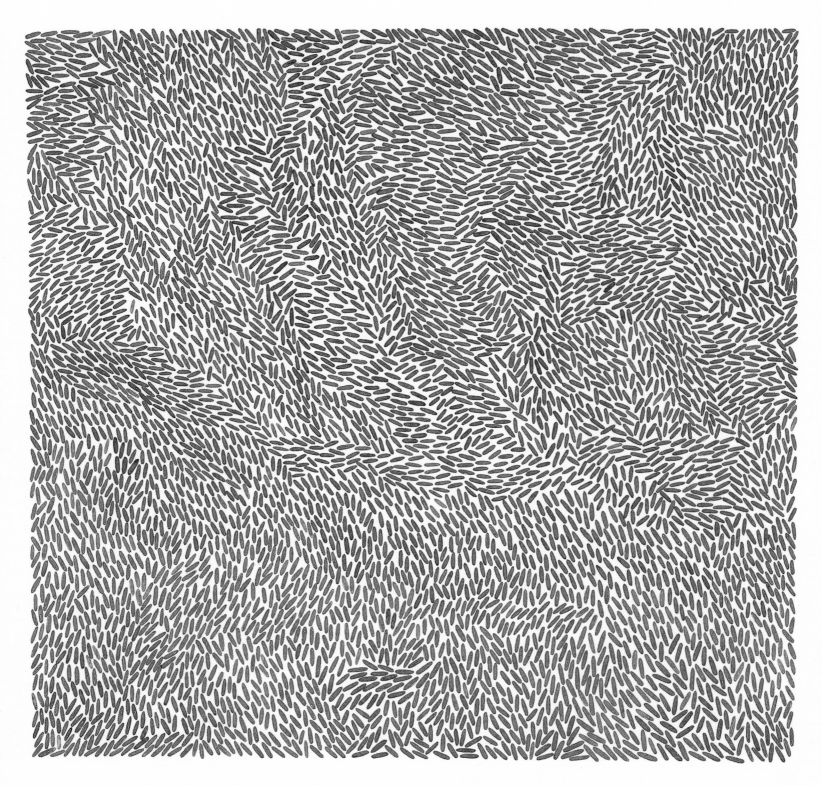

to fill this space, and twenty minutes later ...

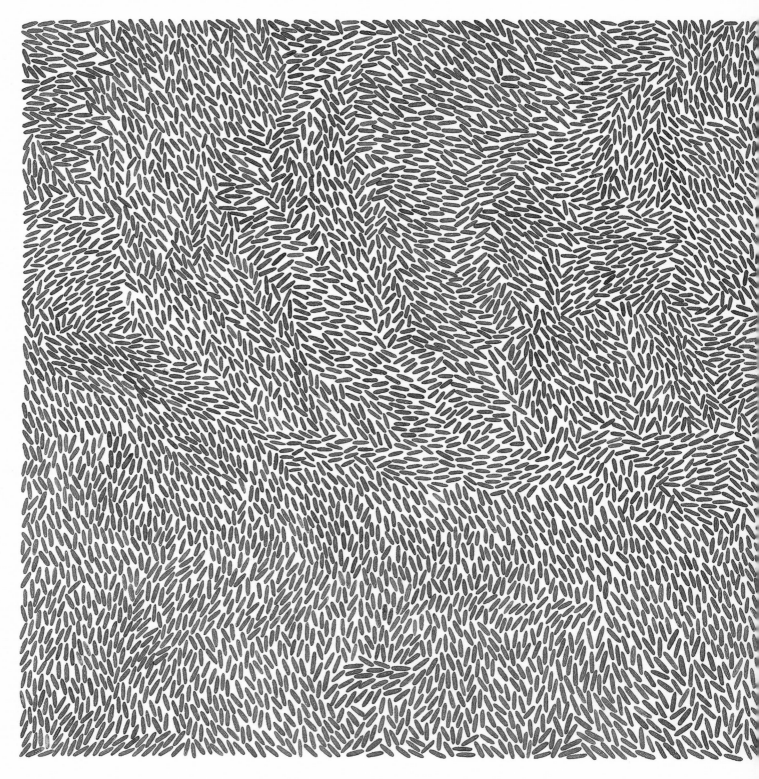

they would all split and

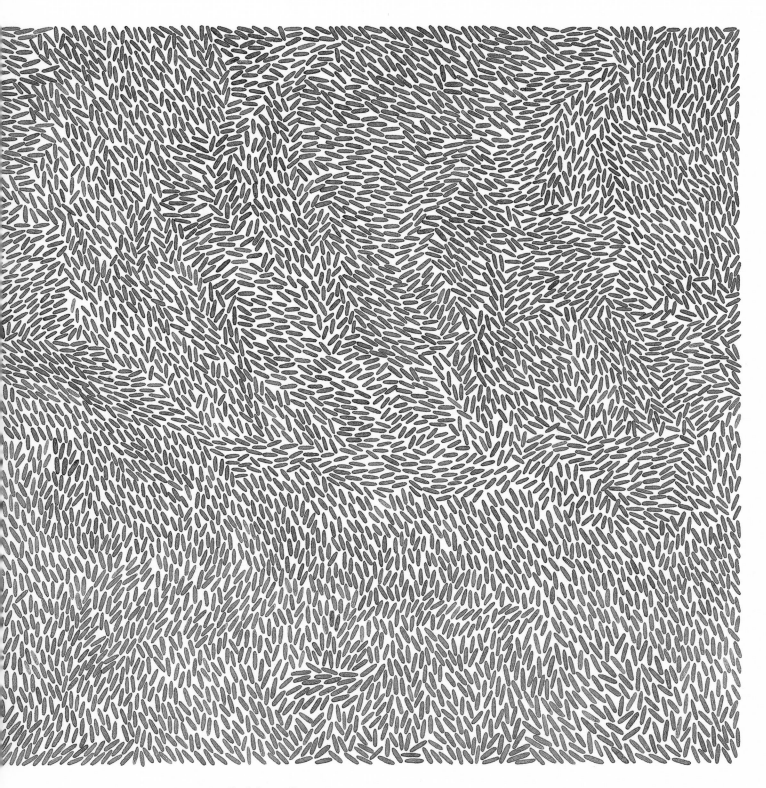

double, to fill this space.

That's why it only takes a few
of the wrong kind of microbes –
the kind we call germs –
to get into your body to make you sick.

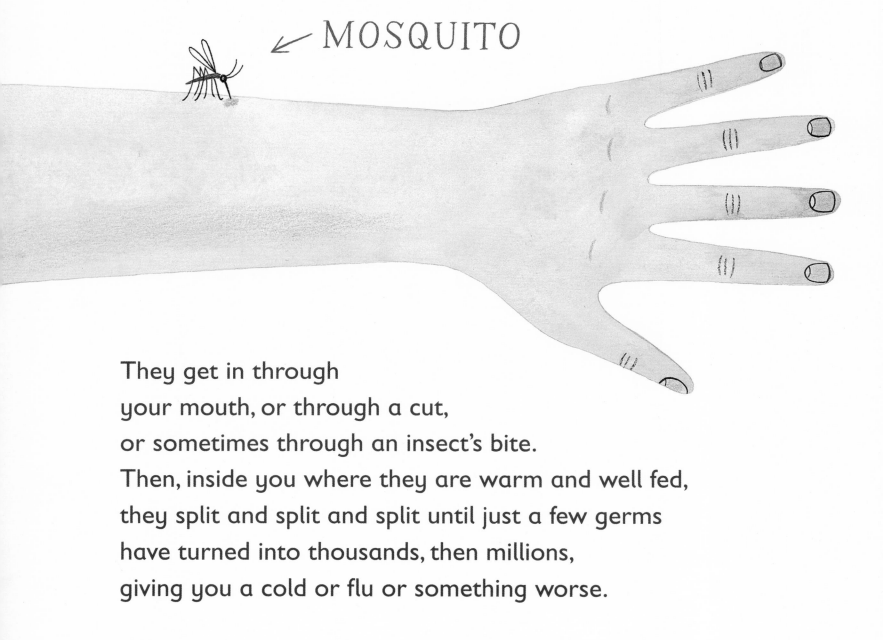

← MOSQUITO

They get in through
your mouth, or through a cut,
or sometimes through an insect's bite.
Then, inside you where they are warm and well fed,
they split and split and split until just a few germs
have turned into thousands, then millions,
giving you a cold or flu or something worse.

So it's best to stop them getting in.

Luckily, only a few kinds of microbes can make humans sick.
Most microbes are busy doing other things.

And, because microbes are
good at making more microbes,
some of the things they do
are very, very big.

They can wear down mountains
and build up cliffs.
They can stain the sea red,
turn the sky cloudy,
and make snowflakes grow.

They recycle everything that dies to make soil so new life can sprout, and they help to make our air good to breathe.

All over the Earth,
all the time, tiny microbes
are eating and eating,
and splitting and splitting,
changing one thing into another.
They are the invisible
transformers of our world.

**The tiniest lives doing
the biggest jobs.**